FROM THE CRADLE TO THE COALMINE

THE STORY OF CHILDREN IN WELSH MINES

CERI THOMPSON

UNIVERSITY OF WALES PRESS
CARDIFF
2014

www.uwp.co.uk

British Library CiP Data
A catalogue record for this book is available from the British Library.

ISBN 978-1-78316-054-9
e-ISBN 978-1-78316-055-6

The right of Ceri Thompson to be identified as author of this work has been asserted by him in accordance with sections 77 and 79 of the Copyright, Designs and Patents Act 1988.

The University of Wales Press acknowledges the financial support of the Welsh Books Council.

Designed by Chris Bell, cbdesign
Printed by CPI Antony Rowe, Chippenham, Wiltshire

He was a miniature miner ... His lips still retained the pout that must have been there when his mother called him before five o'clock, and he kicked his way along as if he hated everybody and everything.

B. L. Coombes, *I am a Miner* (1939)

ACKNOWLEDGEMENTS I would like to thank all the former mineworkers who agreed to be interviewed as part of the Big Pit National Coal Museum's 'People's History Project'. Parts of their interviews appear in this book, but their full stories can be read in Big Pit's annual publication *GLO/COAL*. I would particularly like to thank George Brinley Evans (Banwen Colliery), Bill Richards (Cambrian Colliery), Ray Lawrence (South Celynen Colliery), Arthur Lewis OBE (colliery manager and mining lecturer), Gareth Salway (Bristol City Museums Service) and all the staff at Big Pit National Coal Museum.

CONTENTS

LIST OF ILLUSTRATIONS

CHRONOLOGY OF IMPORTANT DATES

1833 Factory Act introduced to regulate the labour of children and young persons in the mills and factories of Great Britain. The Act does not cover coal mines.

1839 The 'South Shields Committee' set up to investigate the causes and means of prevention of accidents in mines.

1840 A Royal Commission set up to investigate the conditions under which children worked in the mines.

1842 Mines Act: women and girls, and boys under the age of ten, were not allowed to work underground. Boys under the age of fifteen were not allowed to work machinery.

1850 Mines inspectors appointed by the British government – four inspectors employed to cover all British coalfields.

1855 Additional mines inspectors appointed bringing the total to twelve.

1860 Coal Mines Regulations Act: minimum age for employment underground was raised to twelve years old. However, boys from ten years of age could be employed if they had obtained a certificate of school attendance and could read and write.

1872   Mines Regulation Act prohibits the full-time employment of boys under twelve, and boys under sixteen are not to be employed at night or work more than ten hours a day or fifty hours a week. The same act stipulates half-day schooling for boys aged between ten and thirteen.

1887   Mines Regulation Act sets a legal minimum age of twelve years old in the coal industry. Miners had to have had a minimum of two years' experience underground before being allowed to work on their own.

1909   Working day of eight hours introduced in mines.

1911   Coal Mines Act: boys under fourteen years of age may not be employed underground.

1944   Ministry of Fuel and Power appoints a Technical Advisory Committee to examine coal production. Part of their brief is to look at education and training.

1945   Coal Mines (Training) Regulations. These brought in a minimum period of 264 hours training for new entrants, better certificated education courses and medical examinations for all new entrants. They became fully operative on 1 January 1947.

1956   All new entrants to the coal industry given free helmet, boots and other protective clothing. Replacements would be charged at cost price for boots and quarter cost price for helmets.

1957   The Mines (Employment of Young Persons) Order set the minimum age for boys to be employed underground, unless for training, at sixteen years of age.

1972   Wilberforce Award: adult wages to be paid at eighteen years old in the mining industry. School leaving age was raised to sixteen years.

INTRODUCTION

But the young, young children, O my brothers,
They are weeping bitterly!
They are weeping in the playtime of the others,
In the country of the free.

Elizabeth Barrett, *The Cry of the Children* (1840)

CHILDREN HAVE ALWAYS worked but their work has varied through history according to the society to which they belonged. They are still found in rural areas all over the world helping to care for animals, plant seeds and clear weeds, and pick ripe crops. In the cities they swept roads, sold flowers and foodstuffs. They also were sent up chimneys as 'climbing boys'. Children have also been employed in industry for hundreds of years. By the 1840s around a third of the workforce of a cotton mill or coal mine could be composed of children.

Children's work in the Pembrokeshire coalfield was described in 1603 by George Owen:

> then have they bearers which are boys that bear the coals in fit baskets on their backs, going always stooping by reason of the lowness of the pit, each bearer carrieth this basket six fathoms where, upon a bench of stone he layeth it, where meeteth him another boy with an empty basket, which he giveth him and taketh that which is full of coals and carrieth it as far, where another meeteth him and so till they come under the door where it is lifteth up.

Before the nineteenth century, child labour was not always seen as a bad thing. For example, when Daniel Defoe visited Lancashire during the early eighteenth century for his *A tour thro' the Whole Island of Great Britain* (1724), he saw four-year-olds working in the cotton industry and was pleased that they were gainfully employed. The seventeenth century philosopher John Locke went as far as stating that poor children should be put to work at three years old with a bellyful of daily bread! Childhood was generally seen as a time when skills were learned in preparation for adult employment.

During the nineteenth century attitudes towards child labour began to change. There was much social and public debate over the conditions that children worked under and whether they should be in work at all. So what were the arguments for and against the practice?

Working-class children were regarded as 'little adults' and expected to contribute to their family's income. Parents had worked as children and expected their children to do the same. There were few schools available so work kept children out of mischief. Many industrialists saw the employment of children as a cheap source of labour that kept them competitive. They were regarded as ideal industrial workers, they did as they were told and were unlikely to form trade unions. They were small and nimble so suitable for working amongst machinery in a textile mill or in narrow underground roadways.

However, social reformers and Romantic poets such as Elizabeth Barrett were united in demanding that childhood be a period of innocence protected from the harsh realities of life. Another Romantic poet, Coleridge, pointed out that, even though slavery had been abolished in Britain in 1833, little 'white slaves' still laboured in British industry. Parents who sent their children to work were either regarded as wanting more money to spend on themselves or so poverty stricken that they had to send their infants out to work to make ends meet. Industrialists who employed children were accused of being greedy, grasping ogres whose 'dark satanic mills' relied on poorly paid youngsters in harsh and dangerous conditions, while they themselves lived in luxury.

The arguments led to new legislation, and children began to be protected by law. A Factory Act was introduced in 1833 to regulate the labour of children and young people in the mills and factories of Great Britain. Unfortunately, the Act did not cover the coal industry, but one of the factory commissioners recorded that 'labour, in the worst room, in the worst conducted factory is less hard, less cruel and less demoralising than the best of coal mines'. It was obvious that something had to be done.

CHILD MINERS:
THE 1842 COMMISSION

Oh! Mark yon child, with cheeks so pale,
As if they never felt the gale
That breathes of health and lights the smile;
It tells of naught but lengthened toil.
Its twisted frame and actions rude
Speaks mind and form's decrepitude,
And show that boyhood's hours of joy
Were never known to the Collier Boy.

Robert Lowry, *The Collier Boy* (1839)

See, too, these emerge from the bowels of the earth!
Infants of four and five years of age, many of them girls . . .
Their labour indeed is not severe, for that would be
impossible, but it is passed in darkness and solitude.
Hour after hour elapses and all that reminds the infants
of the world they have quitted is the passage of the coal
wagons, for which they open the air doors of the galleries.

Benjamin Disraeli, 1841

If the Welsh mother knew the dangers which awaited the
physical constitution of her child by its exposure to the foul
air of the colliery at the immature age of five years, no
legislative enactment would be required to limit the age
at which the boy should commence work.

R. H. Franks, *Employment of Children in Mines Report* (1842)

THE PLIGHT OF WOMEN and children underground was championed by Lord Ashley (later the earl of Shaftesbury). Ashley had already fought for the protection of factory children in a campaign which led to the 1833 Factory Act and in 1840 he managed to persuade Parliament to set up a Royal Commission to investigate conditions in the mines.

Four commissioners were appointed to enquire into the age, sex, number and the working conditions of children employed in the mining industry. Evidence was taken from employers, doctors, teachers, clergymen, parents, adult miners and the children themselves. The commissioners found evidence of brutality, accidents, long hours, associated lung diseases and horrific conditions of work. It was the first government report to use pictures, and it deeply shocked the public, who were particularly alarmed by the plight of the young 'trappers' (who opened and closed the ventilation doors underground), the nakedness of males and females working together, and what was seen as the lack of religion or morality among the young workers.

The final report of the commissioners was published in May 1842, in spite of a Home Office attempt to suppress it. Thanks to the immense amount of press coverage surrounding the issue it became a best-seller. It became a major talking point amongst the general public and was the subject of intensive debate in both Houses of Parliament as supporters of the coal owners fought to oppose any changes to the use of child labour in the mines. The resultant proposed Bill asked for the exclusion of all females and boys under thirteen from underground employment. Lord Ashley had in fact wanted to raise the minimum age for miners to be fourteen but decided on thirteen to bring it in line with the 1833 Factory Act.

Public outrage at the commission's findings ensured that the Bill was passed by the House of Commons. However, when the Bill was put before the House of Lords, there was fierce opposition from the coal owners and their supporters. The coal owner, Lord Londonderry, argued that 'after the age of ten the boys do not acquire those habits which are particularly useful to enable them to work in the mines'. It was stated that some seams were so thin that they could only be worked by children. In addition, no one was forced to work underground and it was the parents themselves who wanted their women and children to work. They claimed that it would be the poor who would suffer if the Bill succeeded.

The opposition against the Bill ensured that it was modified so that the proposed minimum age for underground employment for boys had to be lowered from thirteen to ten years old. The modified Bill was passed in August 1842 and, in addition to a ban on all females and boys under ten, it stopped the custom of paying wages in public houses, it ensured that a measure of inspection of mines would take place by the employment of a Commissioner for Mines and that all workmen in charge of machinery had to be fifteen years or older.

The resultant Act did not apply to children or females working on the surface of collieries and it did not restrict working hours, even though many young miners were working much longer hours than their counterparts in mills and factories.

The Act might not have completely eradicated child labour but it was a major leap forward in social improvement and the report that led up to it has left us with an incredible amount of information about coal mining in the mid-nineteenth century.

Welsh mining in the early nineteenth century

Few colliery owners employed more than a hundred workers and the 1842 commissioners estimated that around a third of them were less than eighteen years old. The following three collieries can serve as examples:

Dinas Colliery, parish of Ystradyfodwg, Glamorganshire
Adults: 301; under 18 years: 32; under 13 years: 81

Cwmrhondda Colliery, parish of Llantrisant, Glamorganshire
Adults: 35; under 18 years: 7; under 13 years: 5

Morton Colliery, Ruabon,
Men: 18; boys: 7; women: 2

Some children in the north Wales coalfield started colliery work at five or six years old but these appear to have been rare cases, the majority of mineworkers starting their careers at nine or ten years old. It appears that most parents would have preferred an even later starting age. Thomas Williams of Ruabon claimed (rather confusingly) 'those that can keep them at school will not let them go until they are twelve years old; few can keep them so long; they therefore generally go at ten, but very often at seven'.

In south Wales many children started work underground at an earlier age than anywhere else in Britain. John Jeremiah, a steward of Buttery Hatch Colliery, told the inspector that 'Children are carried down almost from the cradle.' This was an exaggeration but it doesn't appear to be unusual for them to go underground as young as four years old –William Richards, aged seven and a half of Buttery Hatch Colliery claimed that 'I been down about three years. When I first went down I couldn't keep my eyes open; I don't fall asleep now; I smokes my pipe; smokes half a quartern a week.' Others started slightly later, William Smith, a 10-year-old collier, was reported to have 'Worked below four years and a half; works with his father and brother; brother is seven years old.' Girls also started their careers at a young age: 6-year-old Susan Reece of Plymouth Works, Merthyr, told the commissioners that she had 'Been below six or eight months.'

The most usual method of coal getting in this period was by driving tunnels, usually termed 'drifts', 'slants' or 'levels', into a hillside. Where this was the case the young mineworkers could not only walk into their places of work but also leave them quite easily during the shift. Little Susan Reece told the commissioner that she 'Sometimes runs home when lamp is out and am very hungry.' This problem was confirmed by the overseer at the Plymouth Mine who complained that the children frequently ran home and left the ventilation doors that they were supposed to be operating. Where geology dictated that shafts had to be sunk to get at the coal, they were normally quite shallow, less than 300 feet deep. However, a few shafts in the western parts of the south Wales coalfield were a lot deeper: down to 232 yards at St David's Deep Pit and 200 yards at Llansamlet Collieries.

Shaft mines were entered either by ladders attached to the sides or workers were raised and lowered on a rope powered by a horse gin, hand winch, steam engine or water-powered 'balance wheel'. Philip Philip, aged ten, from Brace Colliery, Llanelli, was accustomed to the dangers of these primitive shaft mines: 'I help my brother to cart. I can go down the ladders by myself. I am not afraid to go down the pit.' The inspector who interviewed him for the 1842 report climbed down these same ladders with difficulty. Unlike Philip, he was afraid of the noise and the heavy pumping rods that were very close to the ladders. Others were not as lucky as Philip. In 1839, 12-year-old Ann Jenkins was reported killed after she 'fell into a coal-pit' at Graig Colliery, Pontypridd.

In the early nineteenth century, children performed four main tasks underground: horse drivers (or 'hauliers'), doorkeepers, 'drammers' or helping an adult on the coalface.

'Coming to bank' on a rope

Horse drivers

The duty of the haulier is to drive the horse and tram, or carriage, from the wall-face, where the colliers are picking the coal, to the mouth of the level. He has to look after his horse, feed him in the day, and take him home at night; his occupation requires great agility in the narrow and low-roofed roads; sometimes he is required to stop his tram suddenly – in an instant he is between the rail and the side of the level, and in almost total darkness slips a sprig [a piece of wood also known as a 'sprag'] between the spokes of his tram-wheel, and is back in his place with amazing dexterity; although it must be confessed, with all his activity, he frequently gets crushed. The haulier is generally from 14 to 17 years of age, and his size is a matter of some importance, according to the present height and width of main roads.

As a class these youths have an appearance of greater health than the rest of the collier population (probably from their being more in the fresh air than the others), with fair animal spirits; and on horseback, going to or returning from work, galloping and scrambling over the field or road, bear the aspect of the most healthy and thoughtless of the collier-boys.

R. H. Franks, *Employment of Children in Mines Report* (1842)

However, 10-year-old Philip Davies of Dinas Colliery didn't fit this description: he was described as pale and undernourished in appearance and his clothing was worn and ragged. 'I have been driving horses since I was seven but for one year before that I looked after an air door. I would like to go to school but I am too tired as I work for twelve hours.'

Doorkeepers

In order to ensure that the air current reaches all parts of the underground workings a series of 'air doors' are set up. These have to be opened up and closed one at a time to prevent the air flow short circuiting. Because this takes time it was the practice to employ small children to open and close the doors to allow the passage of horses and drams.

The Dowlais Iron Works were the largest in the world at this time and supplied products to many parts of the world. However, they still relied on children to operate the ventilation systems of their collieries. Three sisters, Elizabeth Williams, aged ten, Mary Enoch aged eleven and Rachel Enoch aged twelve, worked in one of their coal mines: 'We are doorkeepers in the four-foot level. We leave the house before six each morning and are in the level until seven o'clock and sometimes later. We get 2d a day and our light costs us 2½d a week. Rachel was in a day school and she can read a little. She was run over by a dram a while ago and was home ill a long time, but she has got over it.'

Phillip Phillips was a door boy in the Plymouth Mines, Merthyr Tydfil: 'I started work when I was seven. I get very tired sitting in the dark by the door so I go to sleep. Sometimes when I am hungry I run home for some bread and cheese. Nearly a year ago there was an accident and most of us were burned. I was carried home by a man. It hurt very much because all the skin was burnt off my face. I couldn't work for six months.'

Doorkeeper Mary Davis was a 'pretty little girl' of six years old. The government inspector found her fast asleep against a large stone underground in the Plymouth Mines, Merthyr. After being wakened she said, 'I went to sleep because my lamp had gone out for want of oil. I was frightened for someone had stolen my bread and cheese. I think it was the rats.' Susan Reece, also six years of age, was a doorkeeper in the same colliery as well: 'I have been below six or eight months and I don't like it much. I come here at six in the morning and leave at six at night. When my lamp goes out, or I am hungry, I run home. I haven't been hurt yet.'

There were conflicting opinions about using children in such employment:

The air-door boy is generally from five to eleven years of age: his post is in the mine at the side of the air-door, and his business is to open it for the haulier, with his horse and tram to pass, and then close the door after them. In some pits the situation of these poor things is distressing. With his solitary candle, cramped with cold, wet, and not half fed, the poor child, deprived of light and air, passes his silent day: his or her wages 6d. to 8d. per day. Surely one would suppose nothing but hard poverty could induce a parent so to sacrifice the physical and moral existence of his child! Yet I have found such to be the case, arising as frequently from the cupidity as from the poverty of parents.

R. H. Franks, *Employment of Children in Mines Report* (1842)

The marquis of Londonderry, a prominent mine owner, gave a different view to the House of Lords:

> The trappers' [doorkeepers'] employment is neither cheer-less, dull nor stupefying; nor is he, nor can be, kept in soli-tude and darkness all the time he is in the pit. The working trap-doors are all placed in the principal passages, leading from the bottom of the pit to the various works, so that an interval of seldom more than five minutes, but generally much less, passes without some person passing through his door, and having a word with the trapper. Neither is the trapper deprived of light by any means generally, as the stationary lights on the rolly and tram ways are frequently placed near the trapper's seat.
>
> The trapper is generally cheerful and contented, and to be found, like other children of his age, occupied with some

Below: *Child sitting by air door*

Sled drawn by boy using a girdle and chain.
Children's Employment Commission

childish amusement such as cutting sticks, making models of windmills, wagons, etc. and frequently drawing figures with chalk on his door, modelling figures of men and animals in clay, etc.

Drammers

The sheer hard work of moving coal underground before mechanisation is difficult to imagine today. The access tunnels leading to and from the workplaces were dark, low and uneven. The male and female haulage workers, often very young, carried baskets or pushed and pulled wooden sleds loaded with coal along these early underground roadways for about the same wages as unskilled farm labourers.

The low roofs of these early mines made young boys ideal for transporting the baskets and twelve yards was probably the limit

'WATCYN WYN' (WATKIN HEZEKIAH WILLIAMS) 1844–1905

Hezekiah Williams (1844–1905) was a noted hymn writer from Ammanford who started work at Tri Gloyn Level where he was paid 8p a day. He wrote a Welsh-language poem describing his thoughts as a 10-year-old pushing and pulling coal carts deep underground. He became a schoolteacher in 1880.

Cart, cart, cart
Before the break of day,
Cart, cart, cart
When night has come to stay,
Cart, cart, cart,
The cart is never out of sight,
In dreams throughout the fitful night,
I push the cart with all my might,
Cart, cart, cart.

Child hauling sled underground

Boy guiding sled down an incline

Boy and sled 'in the narrow veins of Monmouthshire'

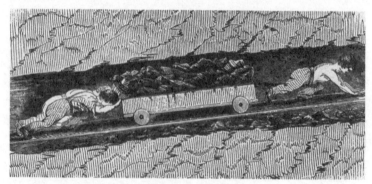

Wheeled cart being drawn by girdle and chain with a little help from behind. Children's Employment Commission

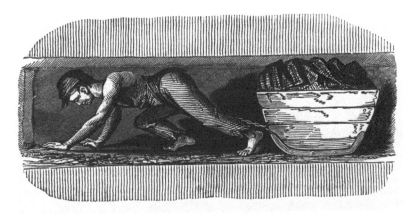

Boy drawing sled by girdle harness

that they could manage with a loaded basket on their backs. It must have been a relief to carry the empty basket back and a chance to gather his strength for the next full load.

Coal sleds were far more common in the rest of Wales than baskets; some of them were even wheeled and ran on flanged rails. The latter were apparently installed in the Neath mines of Sir Humphrey Mackworth as early as 1698. In Mackworth's mines the transport workers were called 'waggoners' and they needed a 'great skill to keep the wagons upon the rails through the turnings and windings underground which are so intricate'. Their skill was so great that they were thought more highly of than the colliers on the coalfaces. However, the job was considered so difficult that few wanted to undertake it.

By the early nineteenth century these 'carters' or 'drammers' were

> employed to drag the carts or skips of coal from the working to the main roads. In this mode of labour the leather girdle passes around the body, and the chain is, between the legs, attached to the cart, and the lads drag on all-fours.
>
> R. H. Franks, *Employment of Children in Mines Report* (1842)

The carts weighed about 1½ cwt of coal and had to be dragged or pushed long distances in low heights. Nine-year-old Edward Edwards of Ysken Colliery, Briton Ferry described his work:

> My employment is to cart coals from the head to the main road; the distance is 60 yards; there are no wheels to the carts; I push them before me; sometimes I drag them, as the cart sometimes is pulled on us, and we get crushed often.

'The collier's little helper'

In some south Wales collieries it was the custom for a child to be taken into the mine by their fathers to claim an extra dram to fill. This enabled a father to earn more and was claimed to be essential for families with children during times of depression in the industry. When the eldest son of a family reached seventeen he also had the right to take a younger brother down. In practice they appear to have only carried out light work doing 'little more than pick up a few coals in loading the carts, and handing and looking after the father's tools'. John Evans was an 8-year-old collier's helper in Gwrhay Mine: 'I've been down two years. Father took me down to claim a dram. I often fall asleep and father pulls me up when he wants me.'

Eight-year-old John Reece worked alongside his father on a coalface in the Plymouth Mines, Merthyr, and seems to have played a more active role: 'I help my father and I have been working here for twelve months. I carry his tools for him and fill the drams with the coal he has cut or blasted down.' Richard Hutton, also of Gwrhay Mine, Monmouthshire, was a year younger:

Opposite: *Woman and young miner in working clothes, mid-nineteenth century*

Dowlais 'tip girl' c.1860. Girls working underground twenty years earlier would have dressed similarly to this.

'I have worked here for a year. The place is middling [reasonably good]. I'm glad when I get home. The shooting [when explosives were used to bring down the coal] used to frighten me and I still don't like it.'

Colliery management informed the commissioners that their employees worked from eight to ten hours a day. However, both adult and child witnesses declared that they never worked less than twelve hours, from six in the morning until six in the evening, and occasionally worked up to eighteen! Whatever the truth, the young drammers and doorkeepers tended to spend longer hours underground than the adults they worked with. The reason for this is that they had to clear the coal cut by the colliers after the latter had finished for the day.

Pay and conditions

The children's wage rates varied between collieries and their pay rose as they grew older and stronger and took over other jobs. For example, in Blaenafon, doorkeepers earned from 10s. to 12s. a month but could earn that in a week when they became hauliers. In Brymbo Colliery, Wrexham, 15-year-old Samuel Thomas earned 1s. 6d a day as a haulier, while in Dinas Colliery, Rhondda, 13-year-old David Morgan could earn up to 9s. a week.

In the Pontypool district boys from seven to ten years old could expect to earn between 3s. and 4s. 6d a week; from ten to fifteen years of age, from 5s. to 12s.; from fifteen to eighteen years of age, between 12s. to 18s. per week, according to the work that they did. Girls got rather less, for example from the age of ten to thirteen years old they could expect to earn between 4s. and 5s. a week, and from thirteen to eighteen between 6s. and 12s. per week, again dependent on the type and amount of work that they did.

Their wages were not usually paid directly to them: 'the collier boy is, to all intents and purposes, the property of his father until he attains the age of seventeen or marries; his father receives his wages, whether he be an air-door boy of five years of age or a haulier of fifteen'.

It appears to have been the custom for most British mineworkers to have breakfast before leaving for work, which might be as early as three or four o'clock in the morning, and not eat again until sometime around noon. Very few British coalfields had fixed meal times during which all work stopped and Wales followed this trend,

> The children and young persons employed in collieries generally take to their work bread and cheese for their meal in the day-time, for which, however, no fixed time is set aside. The haulier eats his food as he drives his horse along; the little air-door boy may take his meal when he pleases; and as the colliers are paid per ton for their work, they too choose their own time. A supper, however, is generally provided for the collier's return, of bacon and vegetables most usually, for the colliers rarely eat much fresh meat during the week.

Morgan Davies, aged seven, an air-door boy in the Dowlais Collieries, described his eating arrangements: 'I take bread and cheese and bread and butter with me and eat it when I want it. I eat it sometimes in the morning and then have none all day. The rats run away with my bag sometimes.'

At this time the collieries tended to only stop on Sundays and for the principal holy days and festivals such as Christmas and Easter. 'There are but few holidays at the works in the neighbourhood of Pontypool, and none that I am aware of for the purpose of allowing

Opposite: *Welsh colliery worker,* c.*1860*

the children recreation and play.' Much of the public outcry against child labour in mines stemmed from the commissioners' report's description of half-naked girls and boys working together. However, this doesn't appear to have been the case in Wales:

> In general, the Welsh women are remarkable for attention to warm clothing, which they secure for themselves in woollens, flannels &c.; nor are they less anxious for their husbands and children – the men and children are always well defended against the general inclemency of the mountain country.

There is little evidence of deliberate and continuous mistreatment of child miners in south Wales, however it did occur and it was more likely to have been their fellow workers doing the beating in order to make them work harder. In the Dowlais Collieries a 10-year-old door boy, Thomas Jenkins, reported that 'The trammers beat him and the others (horse drivers) with a whip when they do not mind to get the coal out for them.' The situation was similar in north Wales: when 16-year-old Edward Price of Broncysyllte was asked whether he had been mistreated underground, he replied 'Yes, the big boys would thrash the little ones; and the charter masters would often beat the boys.' When questioned about the colliery manager's reactions to the beatings he answered 'I don't know; we never made any complaints.'

The most prevalent diseases amongst all coal mine workers were those of the respiratory organs such as bronchitis and asthma. It does not appear that it was recognised that the coal dust itself was causing the problems. James Probert, a surgeon at the Plymouth works in Merthyr Tydfil, reported that

> The employment of children in mines at a very early age tends to produce disease, by exposing a constitution not

matured to foul air; but other causes contribute to this effect. Such children are very much exposed to wet and cold, especially during winter and the rainy season. They are moreover deprived of solar light, which is as necessary to the proper development of animals as vegetables.

He was supported by Thomas Felton, a surgeon in Blackwood: 'The practice of employing children in collieries at an early age, and exposing their constitution to foul air and absence of light, has a tendency to produce diseases of the lungs and eventually to shorten life.'

A coal mine is a dangerous place for adults, so it is no surprise that children were frequently injured underground. The 1842 report mentions some typical accidents.

Henrietta Franklin of Cyfarthfa Collieries, aged eleven, crushed by horse and tram. Not killed.

'I got my head crushed a short time since by a piece of roof falling.' William Skidmore, aged eight, Buttery Hatch Colliery, Mynydd Islwyn

The above children survived but others were not so lucky. In 1801 a young girl of fifteen was killed in an explosion at the Plymouth Levels, while in 1837 a 9-year-old boy was burnt to death in the Cyfarthfa Levels. Also during 1837 the Plas yr Argoed Colliery, Mold flooded in 1837 killing twenty-one people. Eight were under sixteen years old, including Daniel Owen the Welsh-language novelist's father and two of his brothers, James aged twenty-one and Robert aged eleven years old.

There was considerable concern about the short-term effects of the 1842 Act. There was no compensation for the women and boys who lost their jobs because of the Act, and caused much hardship among families relying on the income from their labour.

Hugh Seymour Tremenheere, 1843–9,
the first chief inspector of mines

Alternative work in agriculture or domestic service paid not only lower wages but allowed less leisure time. Charlotte Chiles preferred her present occupation at Graig Colliery, Merthyr to her previous job as a Carmarthen kitchen maid: 'I prefer this work as it is not so confining and I get more money . . . The work, though very hard, I care nothing for as I have good health and strength.'

It was difficult to enforce the Act because of the isolated locations of many mines and the attitudes of both mine owners and some of their workers. In addition, evasion was made easy because there was only one commissioner to cover the whole of Britain and he always gave prior notice before visiting collieries. When the commissioner, Seymour Tremenheere, visited Wales in 1845, he found that 'the main provisions of the Act had not been adhered to'. Boys were being employed without providing proof of their actual ages and, as he had to give prior notice of inspection, it was all too easy for the colliery management to remove women and children from the works until he had gone.

Therefore, many females and underage boys carried on working, albeit illegally, for some years, their presence only being revealed when they were killed or injured: in 1845, a 7-year-old boy was killed in Charles Pit, Llansamlet and another in Eaglesbush Colliery, Neath. Both were burnt to death. However, by the mid-1850s the practice had largely died out in the older mining areas and was not generally adopted in the newer districts such as the Cynon and Rhondda valleys. By the 1860s the attitudes to child labour had changed for the better and much of the pressure to raise the age of employment to twelve years old appears to have come from the miners themselves.

'COLLIER BOYS': THE COAL BOOM, 1850S TO THE 1920S

My mother put my food in a small tin box and filled a tin 'jack' with cold tea, and said 'May the Lord bring you safe home!' as I left the house.

Joseph Keating, *My Struggle for Life* (1916)

*My forefathers were schoolboys on the Friday afternoon
and collier boys on the Monday morning.*

Bill Richards, Cambrian Colliery

THE EFFECTS OF THE 1842 Act reduced the number of children up to the age of ten working underground. However, as the average age at which mineworkers began their career was calculated as nine in 1841, the effect of the Act on juvenile employment was not great. A more significant decline in the employment of young children occurred after the 1872 Mines Regulation Act, which prohibited the full-time employment of boys under twelve. The same act stipulated half-day schooling for boys aged between ten and thirteen which reduced the attraction of mine owners to employ them. In addition, although it did not make school attendance compulsory, the Forster Education Act of 1870 obliged local authorities to provide schooling for the same age group, which also lowered the percentage of young boys in the industry. Between 1842 and 1911 the percentage of mineworkers between the ages of five and fourteen years of age had fallen to 4 per cent of the total workforce.

There were other factors in the reduction of employment of young boys. Technical advances in mine ventilation and underground transport gradually reduced the need for door boys and drammers. In the Cynon Valley, in 1897, following a dispute between men and management at the Nixon Navigation Collieries, it was agreed that the job of door boy be discontinued completely. The hauliers would henceforth open the doors themselves for which they would be paid sixpence. In spite of their reduced numbers, 'collier boys' played an important part in the coal industry throughout the nineteenth century and beyond.

Welsh coal-mining villages of the late nineteenth/early twentieth centuries differed from other industrial communities where there could be a considerable variety of trades and classes. In the Welsh coalfields, for example, mining was the principal (if not the only) industry. For this reason the sons of miners adopted their father's occupation and, unlike other skilled trades, labour in the mines was recruited mainly from the children and often grandchildren and great grandchildren of miners. Therefore, a coal mine was usually the nearest place of work for a miner's child and it was usually better paid than other local work, if it was obtainable. In such a community the vast majority of miners accepted the need for young labour and wanted their sons to enter the industry as soon as possible. This was probably due to financial reasons, but there was always the idea that a boy needed to become accustomed to colliery work as young as possible.

> Well my family didn't mind me going to work underground. It was the only work that was going then; it was either taking a job in the shop for about 5 shillings a week delivering or going down the colliery. So when I was fourteen I went into the colliery. I think I got about 13 shillings, something like that.
>
> Thomas John Williams, Tylorstown Colliery, b.1916

Many young miners couldn't wait to start:

> There is a charm about mining that constitutes a considerable attraction for boys. Underground work appeals to their love of mystery and adventure, and school-boys in mining villages look forward for many years with great eagerness to the time when they will be privileged to descend the pit shaft and experience for themselves the novelty and adventure of which their older companions often speak.
>
> H. Stanley Jevons, *The British Coal Trade* (1915)

As the cage surfaced, I heard . . . a roar of welcome from
a crowd of children gathered to welcome me, amongst whom
were my sisters, Nancy and Olive and my brother George.
I walked to the lamp room with this gang of children;
I was only a child myself at fourteen years of age . . .
My life as a miner had begun.

Edwin Greening, *From Aberdare to Albacete* (2006)

*Edwin Greening in 1927. He later went on to fight for the
Republican cause in the Spanish Civil War*

All the boys in school looked forward with longing to the day when they would be allowed to be in work. Release from the boredom of school might have influenced them but my happiness was not so much in leaving school as in the idea of actually going to work underground. We saw the pit boys coming home in their black clothes, with black hands and faces, carrying their food boxes, drinking tins and gauze lamps. They adopted an air of superiority to mere schoolboys.

Joseph Keating, *My Struggle for Life* (1916)

Some were so eager that their parents lied to get them employed:

As I was only eleven and born in 1873 I would not be allowed to go down the pit for another year, so mother got a pen and, perhaps clumsily, altered 1873 on the Certificate into 1872. On this forged Certificate I got a job at a small pit with a Cornish miner, who ate Cornish pasties in the pit and drank Welsh beer in large lots when out of it. I was paid six shillings per week. Yet even this small sum relieved the pressure on our wants in the home. I worked for about four weeks, then the manager of the pit got hold of me and very gently and sympathetically told me not to come to the pit again. Our bubble was burst; both mother and I were discovered and lost. Nothing further however was heard of the forgery.

Anthony Mor-O'Brien (ed.),
The Autobiography of Edmund Stonelake (1981)

On the 28 January 1860, a boy just under ten years old tried to start work in Gadlys Colliery. Because he had no school-leaving certificate the colliery overlooker had refused to take him on. However, he tried again the next day, actually entering the

'SURFACE GIRLS'

Although females had been banned from working under-ground in 1842, females could legally continue to work above ground. In 1893, 16-year-old Margaret Evans was working in the Llwynypia Colliery brickworks when she became entangled in machinery. The Pontypridd and District Herald reported her injuries in the following sensitive manner: 'Fortunately she escaped with no other injuries other than the loss of her right arm, and this limb had been literally torn off.' Margaret survived the ordeal and became the mother of six children.

colliery with a haulier to ask one of the colliers to take him on. He got separated from the haulier and wandered into old workings where his candle fired some gas which had accumulated there and killed him.

On June 21st 1865, Thomas Rees, an eight years old collier, was killed when he fell off the carriage taking him down Cyfarthfa Colliery, Merthyr Tydfil. The district Mines Inspector, Thomas Wales, tried to fine the colliery manager for employing an underage worker. However, the manager produced a certificate 'purporting to be from deceased's father to the effect that deceased was more than 10 years of age' and no fine was imposed.

HM Inspectors of Mines Report (1865)

Postcard showing a 'collier boy', early twentieth century

This eagerness didn't usually last:

> We were anxious to go, I can tell you! I was so excited
> about it, I didn't sleep the night before, I wanted to know
> what time it was, I wanted to get up. But I've changed my
> tune since then though! Oh aye! . . . Often, after we came
> home from work as colliers' helpers, we used to fall to sleep
> over our dinner, because we were so exhausted. A boy of
> fourteen, fourteen and a half, was expected to do the same
> work as his butty, who was a fully matured man.
>
> Rhymney collier, b.1881

> I know when I was a growing lad after I got home in the
> night and after getting my food . . . feeling too tired and
> stiff and lifeless to get a bath . . . In the morning when
> I was hauled out of bed, I felt it was like going to the gal-
> lows to get up at all . . .
>
> Vernon Hartshorn, quoted in the *Commission
> on the Coal Industry* (1919)

Some started work without even informing their parents.

> My future husband Ronald was fourteen years old on 12th
> December 1931 and therefore old enough to finally leave
> the British school in Talywaun. That same morning he and
> a friend walked over two miles from his home in Abersy-
> chan across the hill to Tirpentwys Colliery to ask for a job.
> The manager took them on straight away. They were given
> a lamp each and sent down the pit. They hadn't expected
> this so had no food or water with them. Luckily, their new
> workmates shared their sandwiches and bottles of water
> with them and they worked their first shift. When Ron-
> ald eventually arrived back home he found the house in

an uproar. His mother was crying because she didn't have a clue where he was and had sent his brothers running around everywhere looking for him. When he did turn up he didn't understand what all the fuss was about and why he was getting a row off his mother. Sadly, the boy who signed on with him was killed underground eight months later.

Mrs Annie Felton

Some parents were dismayed by the news that their son had chosen a career underground.

To my son Alun

Take courage son! 'Tis hard I know
To sweat and labour deep below.
Down in the pit when still so young
With your sums' table on your tongue.

No 'higher grades' for you my boy
You were not born for play, for joy,
But thus to shoulder best you can
The many worries of a man.

Your strength for growth in work to shed
To help us get our daily bread,
Which, when I eat it, bitter thoughts
Does find it sticking in the throat.

But courage son! Heed what I say,
A question will be asked some day,
Why were you, too, forced to become
A party to my martyrdom.

'Huw Menai', 1886–1961

Huw Menai (Huw Owen Williams) wrote this poem after his son had taken a job as a collier's boy in Britannic Colliery, Gilfach Goch, against his parents' wishes. His son Alun Menai Williams wanted to help out with the family's income after his father had been black-listed by the colliery management due to his political activities. Alun also became a political activist and, as a medic, joined the Republican cause during the Spanish Civil War. He died in July 2006, just days before he was due to speak at an event at Big Pit commemorating the Welsh miners' role in the Spanish Republic's fight against fascism. He was the last surviving Welsh miner who had fought in Spain.

Young miner hauling a sled through a narrow access roadway,
early twentieth century

Getting to work

The journey to work might be a fairly short walk for some. However, for others, like the 14-year-old future miners' leader, Will Paynter, it could be several miles.

> There were no buses or other transport to take us the two or so miles from Trebanog to the Coedely Pit. On the morning shift we were raised from bed at about 4:20 am to dress and walk to the pit, collect the pit lamp and be down the pit before 6 am.
>
> Will Paynter, *My Generation* (1972)

'That first descent'

> That first descent into Navigation coal pit, at half past six o'clock, on the morning of my twelfth birthday, April 16th, 1883, interested me wonderfully. As we dropped below the brink of the shaft, the pale daylight seemed to spring upwards and vanish like a flying ghost. For a moment after that I could see nothing at all. Then faint yellow rays appeared from our lamps, and I could see as well as feel the forms of men and boys with me.
>
> We were going down so rapidly – the pit was a quarter of a mile deep – that our lamplight seemed to me to be always running up.
>
> All the time a terrific wind kept shrieking and blowing bits of coal into our faces. The tiny, flying things struck my forehead and cheeks sharply and painfully. I felt as if I were falling through the earth.
>
> The quarter of a minute which was all the time taken for the quarter of a mile drop, had been a magical period in which I had passed from happiness to terror, and back again from terror to happiness. I was delighted to be in the pit.
>
> Joseph Keating, *My Struggle for Life* (1916)

Yr wyf i little collier
Yn gweithio underground
The rope will never torri
When I go up and down
Bara menyn when I'm hungry
Cwrw when I'm dry
Gwely when I'm tired
And nefoedd when I die

Anon (probably late nineteenth century)

yr wyf i = I am
yn gweithio = working
torri = break
bara menyn = bread and butter
cwrw = beer
gwely = bed
nefoedd = heaven

Now here I am, first time to go down a pit and a very deep one at that, stepping into a cage and I used to be too frightened to go on the Big Dipper at Barry Island! I hung onto the side rail like a man clings to a sinking boat. The cage gently rose a few inches and then . . . all hell let loose, my legs parted from my body and it was like I was being shot from a gun down a never-ending black hole. My lamp, which was hooked onto my waist belt never even flickered. Then this mad beast, which a few moments before had seemingly been hell bent on destruction, very slowly and with feather-like softness touched the pit bottom.

Alf Parker, Penallta Colliery

When we got into the cage us boys were put into the centre with the men standing on either side of us. When you went down in the cage it would take your breath away. Goodness knows what it would have been like when they were winding coal as they wound faster then. When you reached pit bottom most of the night shift were there waiting to go up. They never wasted any time winding as they had only half an hour to get all the men down.

Doug Evans, Penallta Colliery

Work

In other parts of Britain, boys began their careers as haulage workers and graduated to working on the coalface. However, in south Wales, young miners became either coalface workers or haulage workers and stayed in that job during their careers in the collieries. There was no official training until well into the twentieth century and they had to learn the job from the adult they worked with (usually referred to as his 'collier' or 'butty'), who could often be his father or other close relative.

I always got on extremely well with my father as a 'butty' . . . (we) worked together with little friction for six years until it was time for me to take charge of a place on my own.

Robert Morgan, My Lamp Still Burns (1981)

Others were not as lucky as Robert Morgan!

> Fathers be the worst butties going. They do think their own
> sons be bloody slaves.
>
> Lewis Jones, Cwmardy (1937)

By the late 1920s it was less likely that a boy would start work
with a relative. By this time boys were taken on by the colliery and
assigned to an adult collier who would teach him his craft.

> I started at Penallta in 1929 working for a collier called
> Bill Martin. Bill was an easy-going type of man with a
> good sense of humour. He also was a very happy type
> of person to work with. Every task was demonstrated
> with much joviality and made the work seem a sort of
> a game.
>
> Alf Parker, Penallta Colliery

A collier's boy was paid a basic wage (known as 'day wage') by
the company but would also receive a percentage of the collier's
wage ('trumps' or 'helpers' wages'). 'Trumps' was a voluntary
system of payment but was usually adhered to by the collier as he
had benefitted by the custom when he had begun his own mining
career.

Boys were shown how to undertake all the tasks demanded
of the Welsh collier: how to cut wooden roof supports and put them
in position, how to cut the coal, how to drill holes for explosives
and, the most common task for boys, how to use a 'curling box'.

Overleaf: *Young colliers arrive at Pochin Colliery from colliers' train,*
c.1920

The curling box

Until the mid-twentieth century, it was usual for lump coal cut at the face to be loaded into a 'curling box' by the collier's boy who dragged or carried it to be filled into a dram. These curling boxes were eighteen inches to two feet wide, three-sided metal scoops with a handle at each side. They were used to reduce the amount of small coal which was sent up the pit and for which there was no ready market at that time.

> The job of the collier's boy was to gather the coal, cut by his mate, into the curling box and to carry it to the tram . . . The collier's boy had to be careful to pick only the lumps of coal, as the collier was only paid for the large

Filling a dram using a curling box, early twentieth century

Billy Hurland filling a dram with a curling box,
Clog & Legging Level, Pontypool, c.1910

coal in the tram. The tram would only hold up to thirty hundredweight of coal, and when it was full it was the boy's responsibility to mark the tram, with chalk, with the collier's number.

J. Griffiths, quoted in G. A. Hughes (ed.),
Men of No Property (1971)

Men and boys at Resolven in 1910.
Notice the curling box in the foreground.

We worked the 2ft 9in seam and of course you were on your knees practically all day. The collier was cutting the bottom of the coal with his mandrel and all of sudden it would bust and it would be all loose and it would all come out, slide out, and then you'd have to get that coal into the box and I'd be dragging it back up to fill the dram. And I was short then, so we had to have a box at the side of the dram for me to stand on so I could lift it and put the coal in. I was really tired by the time I went home!

Thomas John Williams, Tylorstown Colliery

Using a curling box for the first time was a difficult operation. Most inexperienced mineworkers tend to move around on their knees, which soon become tender and sore. In addition, they had to manoeuvre around the roof supports before lifting the laden box high enough to be poured into the dram. Fourteen-year-old Robert Morgan was taught the correct way by his father.

> It was to move through the face on one's toes, the right hand gripping the handle of the box and the left hand pawing the floor as one moved along . . . It took many weeks before I finally mastered it and, looking back, I must have looked like a swift crab, with my knees almost touching my chin and the pawing hand keeping the ball-like body balanced.

On top of this he found that the dram was too high, and the coal-filled box too heavy, for him to fill it. He had to place a spare sleeper to stand on but, even with this help, he found difficulty.

> But I persevered by first lifting the box to my knees, resting it for a moment, jerking it on to the rim of the dram, and then snatching the box away, the contents falling into the dram.
>
> Robert Morgan, *My Lamp Still Burns* (1981)

The danger in using these boxes can be shown by the death of a young boy in Seven Sisters Colliery in 1921. The boy was filling one of these boxes when a large stone fell from the roof hitting him on the back and forcing his body onto the curling box which cut into him. He died while being carried on a stretcher to hospital.

Because of the unpopularity of the curling box, many mineworkers tried to avoid using them as much as possible.

Men and boys outside Cambrian Colliery lamp room, c.1913

Powell Duffryn insisted that you fill the tram with a curling box to ensure that there was no stone in it. When the
overmen and firemen were about the word would pass
up the face so you had to use the curling box! You had to
scrape it in with your hands, lift the curling box and put
it in the tram. When they had gone we would go back to
using the shovel!

The trams had to be bedded, and then raced-up about two-foot. You put the small coal inside; you had to keep the big lumps for the top. Powell Duffryn only paid the colliers for lump coal. The rest would go to the washery and then be sold after all. I didn't know these things at the time. As you got older the colliers would tell you these things. We'd fill five or six trams of coal a shift, each holding a little more than a ton each.

Doug Evans, Penallta Colliery

Not all young mineworkers went straight underground.

I started work at fourteen years old. I was supposed to work underground but I was put on the screens instead. These were screening 1,000 tons a day and the noise and the dust were horrendous for the whole eight hours shift.

There was no dust suppression in those days and all communication had to be by sign language.

I would have been better off underground!

George Evans, Banwen Colliery

The screens at Court Herbert Colliery, 1913

LOST IN THE MINE

One Friday afternoon, William Withers went to work with his father as usual. On arriving at the mine, he found that he had left his lamp at home and returned to get it intending to catch up with his father underground. However, while walking along the underground road-ways his light went out and he wandered into a disused part of the mine. This is his story:

After I lost my light, I found that I was lost, and in a strange road. I could hear my father at work all Friday, I knocked the side, and made as much noise as I probably could, but no-one answered me. They all went out that night, leaving me there; I cried very much. I thought I saw the stars two or three times, although I was a hundred yards underground. I saved my dinner as much as I could, only eating a bit at a time, not knowing whether I should ever be found. The pit broke (closed) on Saturday morning, so there was no work until Monday morning. The whole time I had been wandering about in the dark, when I heard the hauliers, and made my way to them.

When asked what day he thought it was, the poor little fellow did not know, but thought he had been lost seven or eight days.

'Our coal fields and our coal pits' (1853)

THE DOOR BOYS STRIKE

Young mineworkers took part in all the major and minor disputes that have punctuated the history of the coal industry. However, there have been times when they took action on their own.

In 1864, door boys at Abercwmboi Colliery withdrew their labour until the owner agreed to increase their wages from 7s. a week to the 8s. that their counterparts were earning in other collieries. The owner, David Davis, asked the colliers to perform the boys' jobs as well as their own for which they would receive the boys' wages on top of their own, which the men refused. A compromise was reached after nine weeks and the boys were awarded 7s.8½d, which was the average wage for their district.

Almost a hundred years later, another young collier was not as fortunate. In 1933 the Mines Inspector for the Cardiff District reported that a boy had wandered into a disused roadway which was fenced off because it was known to contain firedamp. When he was missed and a search was made he was discovered dead near the face of the roadway. The unsympathetic opinion of the inspector was that 'He was old enough to appreciate the significance of crossed timbers in any roadway as denoting danger and a probation from entering.'

Death and injury

Because of their inexperience, and sometimes their natural inquisitiveness, young mineworkers were at a greater risk of injury and death than the adults they worked alongside. Between 1851 and 1855, one-fifth of deaths in coal mines in south Wales were boys between ten and fifteen years old; this was in spite of the fact that they were only one-ninth of the workforce. Many of these victims were air-door boys who, instead of being given a proper place beside their air door and orders to stay there, were allowed to run in front of journeys of drams to open the doors for the hauliers. It was all too easy for them to slip and fall under the heavy drams and, as a result, death from 'crushed by dram' was all too frequent.

Collier boy at Bargoed Colliery pit bottom, early twentieth century

The situation was similar for the next hundred years. For example, between 1929 and 1933 16.9 per cent of all miners in the Swansea District of HM Inspectors of Mines under twenty years old were killed or badly injured. The problem was so great that this district saw the formation of classes for boys of fourteen years and upwards in safety principles which were 'taken up with some enthusiasm in some mines'.

However, safety training remained inadequate for many years to come. In 1938, 16-year-old Arthur Lewis was the only first-aid man amongst 350 miners.

Tom, who lived in my street, was aged 14 and had been underground in Llanhilleth Colliery for only two weeks. He was working as a collier's boy and his adult 'butty' had let him stand on a dram of coal to rip the roof top coal down to make room for the horses to pass through. The top coal and part of the stone roof fell onto him. When I was called to the scene I saw that his injuries were very severe. Apart from fractures to his arm and ribs he had a gash the whole width of his forehead which had exposed his skull.

After splinting and bandaging his injuries, I had him taken on a stretcher to the first aid room on the surface where a doctor ordered that he be taken to the Royal Gwent Hospital in Newport. I accompanied him in the ambulance and he was conscious at all times but kept saying 'Thank you' to me all through the journey.

When we arrived at the hospital both of us were still in our working clothes and covered with coal dust. The sister in charge took him away to clean him up. When she came

Opposite: *A young Arthur Lewis in his St John's Ambulance uniform, 1932*

back to me with his clothes she was in tears because Tom was continuously thanking her. I remember her words 'I can't clean him any more as he is still conscious – how old is he?' I replied '14 years old', 'And how old are you?' she asked, '16' I replied. She looked at me and said 'You both should be in school not down the pit.'

Disasters

One of the most tragic aspects of the great colliery disasters of the nineteenth and twentieth centuries was the young age of many of those killed. Between 1851 and 1920 there were forty-eight disasters in the south Wales coalfield, causing some 3,000 deaths; many of these were collier boys.

On 17 August 1849, an explosion at Lletty Shenkin Colliery, Aberdare killed fifty-three miners, including fourteen boys under sixteen years of age – one was ten and another eight years old.

The Cymmer Colliery explosion of 1856 was the first major disaster in the south Wales coalfield: thirty-four of the 114 persons killed were under sixteen years of age. Seven were under ten, seven under eleven, eight under twelve and twelve under sixteen years old.

In the National Colliery explosion of 1905, nearly half the 119 victims were boys under twenty years of age, fourteen of these being under fourteen years old. A local paper reported their deaths:

On the afternoon of Wednesday, 103 bodies had been discovered but not brought from the places where they were found. It was tragic to see the large number of boys from

Opposite: *'Ha'penny' – a collier boy from Tonypandy, c.1930*

DISASTER AT CHRISTMAS

Lewis Davies usually worked alongside his father cutting coal in the 'East Rhondda' underground district of Maerdy Colliery, Rhondda Fach. On 23 December 1885, however, he was helping Richard Lewis, an adult haulier, with his horse and dram. At about 2.40 p.m. they heard the sound of an explosion and turned and ran. As they ran a huge fireball rushed along the roadway behind them and knocked them over into the tangled heap of their horse and dram.

Lewis's father, Morgan, also heard the explosion rip its way through the mine workings. His immediate thought was for his child and he ran into the roadway where he met with fourteen other miners who were making their way to the shaft. Within a short while twelve of them had collapsed and died from the effects of the afterdamp. However, Morgan carried on and, in spite of the lack of oxygen, managed to stagger a distance of about 500 yards to where he hoped his son would be. Suddenly he heard the boy calling out in the darkness for his mother and struggled the last few yards to find Lewis trapped between the badly burned horse and the dram which had overturned on top of Richard Lewis. He managed to free the boy from the harness and carry him to the side of the roadway.

Miraculously, all three were rescued but eighty-one other men and boys died in the disaster.

'Young and old at Ferndale Colliery', 1907

Nantyglo colliery boys and schoolboys, c.1900

Pontypridd colliers, c.1910

Father and two sons at Wattstown, c.1920

fourteen to sixteen years among the victims. These bright cheerful lads, who worked with their elders, sometimes relatives, were found in all sorts of positions. One of them quite a little boy laying as if peacefully asleep. Many of them had come under the influence of the recent religious revival and many who travelled on the workmen's trains were heard to sing the revival hymns on their way to and from work.

Many were so badly injured as to be unrecognisable, 'A fy machgen bach i yw hwna?' (Is that my little boy?), asked one father as a young boy was brought into the blacksmiths' shop which was being used as a makeshift mortuary.

MINING TRAINEES: FROM DEPRESSION TO NATIONALISATION, 1920S TO THE 1980S

They are children of the depression, reared on the dole, thrown on the scrap heap and allowed to rust.

James Griffiths, MP

THE DECLINE

THE COLLAPSE OF THE coal industry between the world wars
turned Wales into an area of mass unemployment. By 1932, 36.5
per cent of the working population was unemployed. In some areas,
especially around the heads of the valleys, the situation was even
worse with up to 75 per cent out of work. Unemployment led to
around 500,000 people, mainly the younger and more able, leaving
the coalfields to look for jobs in the new manufacturing industries
in places such as Swindon, Slough and Oxford. Those that stayed
faced living on benefits and charity or, for those still in work, falling wage levels. Many faced 'short-time' working and might only
work a few shifts a week. School leavers became reluctant to enter
the coal industry. The work was hard and dangerous but the wages
were low on the scale of industrial earnings and there was a high
chance of unemployment caused by the frequent mine closures.
Transport systems had also improved and this encouraged potential miners to obtain jobs outside their communities. In addition
the low birth rates in the mining districts during 1926 and 1928
resulted in a shortage of 14-year-old potential recruits after 1940.
Coal was regarded as a dead-end industry and many boys were dis-
couraged by their parents from entering it.

In 1931, around a third of mineworkers were aged over forty
but by 1945 this percentage had increased to almost 45 per cent.
The coal industry was increasingly relying on men who had passed
their prime. Those youngsters that were employed often had to do
the jobs of men (at a lower rate of pay) who had joined the forces

before the Essential Work Order of 1941 prevented mineworkers from leaving the industry. There was a deep sense of discontent amongst some young miners which was added to by their low rates of pay compared to that of their sisters or girlfriends working in the munitions factories. Bert Coombes describes the attitude of some of them at his colliery:

> They are just a section of our mining youth, not the largest section by any count, but their insolence and indifference to all discipline make them a problem in our work and future . . . What mistake in environment and education has brought these young lads into this condition?

Colliery official addressing young miners underground in 1943

A young trainee miner in the south Wales coalfield

The general disgruntlement led to high absenteeism among the young miners and a series of unofficial strikes, sometimes with the support of older workers. The biggest UK-wide strikes occurred during May and June 1942 when over 10,000 men and boys came out, 7,500 in October 1943 and 100,000 (in south Wales alone) in March 1944. George Brinley Evans of Banwen Colliery took part in one of the smaller, local disputes:

> It was 1942 and the strike lasted only a few weeks, but there were a few hundred boys involved and was very

disruptive. The strike put people to a great deal of trouble and some of us were fined by the 'Labour Officer' who had been brought in during the war to impose fines on any worker he thought was damaging the war effort. My parents thought I deserved it; my father said 'You shouldn't have had so much to say for yourself, should you?' My interview with the Miner's Agent was even worse – 'Do you know there's a war on? Do you think that your Bolshie behaviour is helping the country? You're a disgrace to your family. A mouthy little Bolshie! Get out of my sight! Clear off!' Others thought we should have been shot! I was completely humiliated.

George later joined the army and took part in the liberation of Burma.

In 1943 it was decided that, because of the shortage of coal miners, young men of military age were to be balloted and conscripted into the mines. These young men became known as the 'Bevin Boys', after Ernest Bevin the Minister of Labour who introduced the scheme. Their employment in the mines was often opposed by well-off parents who saw mining work as something that other people's children should do. Their concerns were commented on by Mr G. Jones, secretary of the Warwickshire Miners' Union:

They may ask themselves why they object to their boys going into the mines. Is it because it is dirty and because it is dark and dangerous? If they only object on these grounds, are they only objecting for their own child or do they object for all children? If they say that it is too hard, too dark and too dangerous for their boys of 18 years of age, will they also say that it is too hard, too dark and dangerous for other boys of 14 years of age in the mines

Bevin Boys at Oakdale, 1944

today, and who are compelled, under the pressure of poverty, to go to the mines to help in the getting of coal.

After being selected, a Bevin Boy was sent to a training centre for a month's basic training which included classes in mining and physical exercise before he started work in a production colliery. This was more than the local 14-year-old collier boys had ever received when they entered the industry – they were usually sent straight underground on their first day!

Nationalisation

The coal industry was nationalised on 1 January 1947 and one of the first tasks of the National Coal Board (NCB) was to examine the training of young miners. It was decided that it was no longer good enough to leave instruction to the men they worked with. Training schemes were to be set up and these were to be organised partly on the experience gained in training the Bevin Boys.

Under the new training schemes a boy would be recruited from school, perhaps after a talk by a local colliery training officer. He would then be medically examined and, if proved fit, would report for work at the nearest NCB training centre. Here, he would be instructed in a mock up of underground workings known as a 'mining gallery' under the supervision of instructors with personal experience of mining work. He would also be taught about safety, mining practice, first aid and other mining activities. Sometimes sporting activity took place. When he qualified (and this was apparently not too difficult!) he chose which colliery he wanted to work in and signed on there.

> I went in straight from school in 1974. I finished school on the Friday and I started on the Monday in Britannia mining school. I went over to Penallta, knocked on the training officer's door – Bob Davies. And he said 'Turn up to Britannia on Monday for a medical'. So I went over there on the Monday and they took a busload of us from there to Cwm Medical Centre [Llantwit Fardre]. I passed the medical. Then I had to do a spelling exam. That was it! Spell 'horse', spell 'ambulance' it was! Then a colour eye test. Then you had an eighteen-week course in Britannia.
>
> You'd go on visits to other collieries. In Britannia itself they had an underground gallery where they would teach you how to work safely. I went to Merthyr Vale on a visit where I saw horses underground. At the end of your

eighteen weeks you had an exam and if you passed that you were sent to your local colliery. I think I was the only one to be sent to Penallta in that intake.

Dean Wood, Penallta Colliery

At every NCB colliery there would be a training officer who dealt with any problems that the boy might encounter. Later he might be offered more advanced courses – even up to university level. He remained a trainee in the colliery until he was deemed fit to work on his own. The NCB also introduced a scholarship scheme which it was hoped would attract boys from the grammar and public schools who wouldn't normally consider a career in the coal industry. In addition, the NCB's 'Ladder Plan' was in place. Through this scheme a suitably able and ambitious 15-year-old entrant could rise to some of the highest posts in the coal

NCB recruiting exhibition, Cardiff, 1956

industry. As one of the NCB's divisional chairmen, Sir Ben Smith, pointed out, the industry needed 'virile, intelligent youths' and mining was an industry that 'young lads need not be ashamed to enter . . . Mining afforded those with brains and character excellent opportunities for advancement'.

Although his initial training was streets ahead of what his predecessors had received before the nationalisation of the coal industry, Bill Richards found that it still didn't prepare him well enough for the reality of a production coalface.

Young miners in a training coalface, 1952

My heart sank when, after crawling and dragging ourselves through a forest of closely erected props, my butty, Tom Williams and I reached our workplace. Instantly I was filled with trepidation, for the Deep Seven District of Cambrian Colliery No. 4 pit was completely opposed to the comfortable environment of the National Colliery Training Face.

In my new workplace, hunched over and on my knees, there was barely enough space to wield a shovel, while just a few feet away a section of weak, unsupported roof had a menacing presence. Tom plunged the pneumatic pick into the un-watered coal-face and a huge cloud of thick swirling dust engulfed me and almost obscured the glow of my eleven pound electric hand lamp. Our workplace was so confined that his every shovelful of coal had to be thrown to me and I had to re-shovel it onto the conveyor. All my body orifices were clogged by a fine invasive dust, and the oily vapour from the exhaust of the compressed air powered conveyor; whilst from my fingers to my elbows I bore blisters, grazes, cuts and lumps! How the hell had my harder worked predecessors coped without proper training – they had been schoolboys on Friday and collier boys on Monday?

Allan Price also discovered that colliery work was very different from his experience in the training centre.

I attended Maesteg Training Centre in 1959. I learned all the basic things like how to shovel with both hands, how to put up timber and arches and how to build packs and fill them with rubbish. Also for 2 days a week we would go to visit local collieries and work on the coal face or heading with the men. This was a lot better than the training centre because it showed what the work was really like.

Instructor showing a trainee how to bore a shot firing hole, 1950s

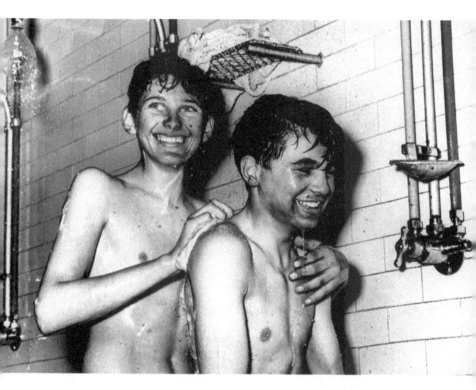

'End of the shift': mining trainees showering in a pithead bath, 1960s

It had been different for my father; he had started at 14 and worked with his 4 'brawds' [brothers]. He told me he started work in the 'muck hole'. What he meant was he was packing the muck off the heading into the gob. When he came home on his first day his 'brawds' let him have the first bath in the tin bath.

These training centres improved their facilities as mining techniques changed over the years.

Mining trainee being taught to cut roof supports with hatchet, 1950s

Trainee being shown coal being loaded from a conveyor into a dram, 1952

Trainees being shown how to set a 'pair of timbers' as roof support, 1952

*Young miners erecting wooden roof supports at Ogmore Vale
Training Centre, 1950s*

Power loading

By the mid-1970s 'Power loading' had become commonplace in the Welsh coalfields. Under this system of winning coal a cutting/loading machine travelled along an armoured flexible conveyor (AFC) which ran the length of the coal face. As well as cutting the coal as it travelled along it also directed the cut coal onto the conveyor. The roof was either supported on bars held up by single hydraulic adjustable steel props or, increasingly, on 'walking chocks', which was an arrangement of two to six hydraulic props grouped on a steel base which held a canopy that could be lowered or heightened against the roof of the seam. Britannia Training Centre, near Bargoed, was equipped with 'chocks' and AFCs which were said to be better than those available in most working collieries.

Andrew Williams (born 1960) had already worked in his family's small mine before undertaking NCB training at Britannia at eighteen years old:

> I'd already spent a year underground on pick and shovel work before visiting Penallta Colliery as part of the training. The first time I saw a 'power loading' face I thought 'Wow!' It was really impressive to see a machine cutting more coal in seconds than I could cut in a whole shift with a 'shaft' [coal pick]. Although I was gobsmacked I also thought that these miners were really spoilt because they only had to move levers on machinery rather than being up to their knees in water cutting coal with a 'shaft' or 'puncher' [pneumatic coal pick] as we had to in the small mine.

Alan Lancaster also found modern mining impressive:

> I started my basic training in 1976 and was taken to see the S10 District in Britannia. It was only a yard high and was supported by hydraulic props and bars rather than

more modern 'walking chocks'. I could see the cutter chain [which enabled the machine to haul itself along the coal face] banging against the roof bars and clouds of dust and then I saw the cutting machine coming down the conveyor towards me – the first time I ever saw a 'trepanner' [a machine that cut out a cylindrical core of coal in the same way as a surgical trepanner removes bone from a skull] a very impressive sight as it busted the coal out from the face and onto the 'panzer' [the armoured flexible conveyor which ran along the coal face].

Mining trainees being issued with protective clothing, 1960s

'WINDING UP'

I was a trainee at Wattstown Training Colliery when an
instructor said to one of us new entrants 'You'd better get
a different shovel butty – that's a left-handed one!'

Bill Richards, Cambrian Colliery

Collieries are famous for 'leg pulling' and naive young miners were
especially fair game. Many boys were sent to obtain tins of striped
paint, glass wedges, sky hooks, left-handed spanners or buckets of
steam or 'blast' (compressed air). One of the guides at Big Pit gained
the nickname 'Bucket' after falling for the last one!

While repairing a machine underground at Roseheyworth Colliery
in the 1980s, a nut was found to be too thick to fit and the fitter's
apprentice was sent out the two miles to the colliery fitting shop on
the surface to modify it. He carried out the task and travelled the two
miles back to the coalface. Unfortunately, instead of cutting the nut
horizontally, he had cut it vertically! He was thereafter always known
as 'Halfnut'!

Trainees being shown how to test for gas

Until the late 1950s, levels of employment in the coal industry remained steady but with a decline in the use of coal and the challenge from oil from the Middle East, there came a period of colliery closures. Between 1960 and 1970 over sixty Welsh collieries closed. The pace of closures slowed as the price of oil increased during the early 1970s and a career in the coal industry began to look more secure. The National Coal Board organised vigorous recruitment drives during this period to ensure that sufficient numbers of young people were attracted to the industry before the school-leaving age was raised to sixteen in the autumn of 1972.

Paul Meredith, now conservation engineer at Big Pit National Coal Museum, remembers that confident period:

> The recruitment drives in the local schools, the propaganda leaflets, training booklets and the talk about hundreds of years of coal reserves, all led us to think that we had a 'job for life'. There were three deep mines around my home town of Abertillery, each employing around 500 men, so my future seemed as secure as my father's and he worked in the pits for 40 years. Britannia Training Centre seemed to be bursting with electrical and mechanical apprentices and mining trainees. There were boys who had got good qualifications in school being 'fast tracked' onto degree courses. In my new nice clean blue overalls and shining white helmet, my future in the industry seemed very bright indeed!

However, within a decade, the industry would experience a further round of closures. The 1970s torrent of new entrants had slowed to a trickle by the mid-1980s and the Welsh coal industry gradually dwindled away to the handful of small mines which exist today. Coal is no longer king and the option of 'going down the pit' is no longer available for most school leavers in the Welsh coalfields. This is a mixed blessing for most mining families:

> I never wanted my son to follow me into the colliery, so in a way I'm not sorry that the pits have gone. However, the pits did create much needed employment and any job is better than none!
>
> Paul Meredith

Previous page: *Trainees enjoying a cup of tea in the canteen*

CHILD MINERS TODAY

*It is unthinkable that so many children are still being
exposed to such acute danger in mines around the world . . .
Urgent action is needed to make sure they are properly
protected from the perils of mining.*

Daniela Reale, Save the Children, 2007

UNICEF (International Labour Organisation Facts on Child Labour) estimates that there are currently around 158 million children aged between five and fourteen years old at work – around one out of every six children in the world. Many of these are employed in mining which is still one of the most deadly forms of child labour. Tens of thousands of children still work the same long hours in similar hazardous conditions as young Welsh children in the nineteenth century. They risk death from explosions, roof falls and machinery accidents. They breathe dust and deadly gases. The reason they work is also the same – poverty!

Modern child miners can be found in over fifty countries around the world. In the Philippines, nearly 18,000 children are involved in gold, silver and copper mining. In the diamond mines of Sierra Leone, child miners, some as young as ten years old, carry bags of gravel, of up to sixty kilograms in weight, on their heads. One of these young mineworkers, Tamba James aged fifteen, lost his parents during the civil war: 'I have to go into the mines because it is the only place I can raise money to feed myself and my brothers.' They regularly work from dawn to dusk, often without proper food and have little medical care if they are injured or fall sick.

In 2005, Mykhaylo Volynets, leader of the Ukraine Confederation of Trade Unions, stated that there were around 6,000 illegal small mines in the Ukraine, many of them family run but some operated by criminal gangs. In these mines many children, some as young as five years of age, work for long hours in unsafe conditions using primitive tools. Effective ventilation, roof support and gas detection are almost completely absent.

Child miners may be a thing of the past in Wales but small hands are still at work underground.

The worst thief is he who steals the playtime of children.

'Big Bill' Haywood, American miners' leader, 1869–1928

FURTHER READING

Arnot, R. Page (1967), *South Wales Miners (1898–1914)*, George Allen & Unwin Ltd.

Arnot, R. Page (1975), *South Wales Miners (1914–1926)*, Cymric Federation Press, Cardiff.

Ashworth, William A. (1986), *History of the British Coal Industry Vol. 5 1946–1982*, Clarendon Press.

Coombes, B. L. (2nd edn 1974), *These Poor Hands*, Victor Gollancz Ltd.

Egan, David (1987), *Coal Society*, Gomer Press.

Evans, R. M. (1972), *Children in the Mines 1840–42*, National Museum of Wales, Cardiff.

Flinn, Michael W. (1984), *History of the British Coal Industry Vol. 2 1700–1830*, Clarendon Press.

Hatcher, John (1993), *History of the British Coal Industry Vol. 1. Before 1700*, Clarendon Press.

Montgomery, George (1994), *A Mining Chronicle*, Newcraighall Heritage Society.

Morgan, Robert (1981), *My Lamp Still Burns*, Gomer Press.

Morris, J. H. and L. J. Williams (1958), *The South Wales Coal Industry*, University of Wales Press.

Reports from Commissioners (1842; repr. 1968), *Children's Employment Commission*, Irish University Press.

Supple, Barry (1987), *History of the British Coal Industry, Vol. 4 (1913–1946)*, Clarendon Press.

INDEX